Photosynthesis

Rebecca Woodbury, Ph.D., M.Ed.

Gravitas Publications Inc.

Photosynthesis

Illustrations: Janet Moneymaker

Photosynthesis
ISBN 978-1-950415-73-1

Published by Gravitas Publications Inc.
Imprint: Real Science-4-Kids
www.gravitaspublications.com
www.realscience4kids.com

RS4K

Photo credits: Cover & Title Pg. joffi from Pixabay; Above & P.7. lovelyday12, AdobeStock; P.15. Vink Fan, AdobeStock; P.17. rustamank, AdobeStock; P.19. Valentin from Pixabay

How do plants eat?

Do plants drink milkshakes?

I don't think so.

Plants make their own food
with **photosynthesis.**

Do you know what that is?

No. But let's find out.

But what is
photosynthesis?

Photo means light.

Synthesis means to make.

Photosynthesis means
to make with light.

Plants use **photosynthesis** to
make their own food with sunlight.

All of the green parts of a plant
use photosynthesis to make food
for the plant.

The green parts of a plant are green because of a special molecule called **chlorophyll.**

Do you think I have chlorophyll?

Duh! Are you green?

Review: ATOMS

Atoms are tiny building blocks that can link together.

Atoms make everything we see, touch, taste, and smell.

Review: MOLECULES

Molecules are made when **atoms link** together.

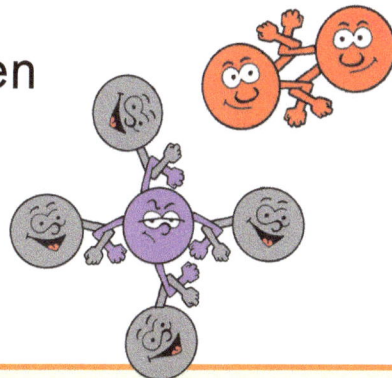

Chlorophyll is a
BIG molecule!

· · · · · · · · · · · · · · · · · ·

Look at all
those atoms!

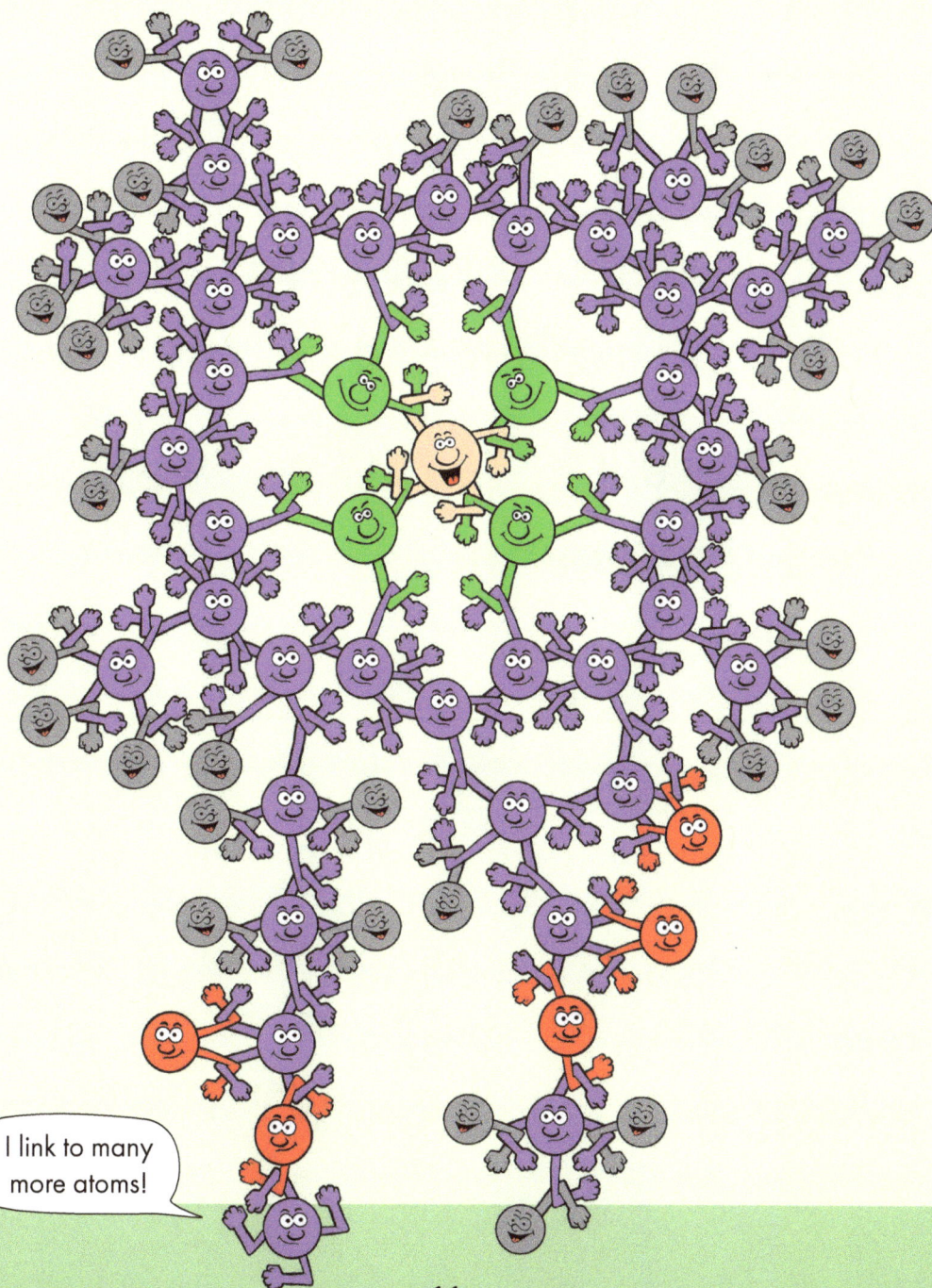

I link to many more atoms!

The job of chlorophyll is to catch sunlight and send it to a sugar factory called the **chloroplast.**

Yay! Sugar!

I don't think this comes in a bag.

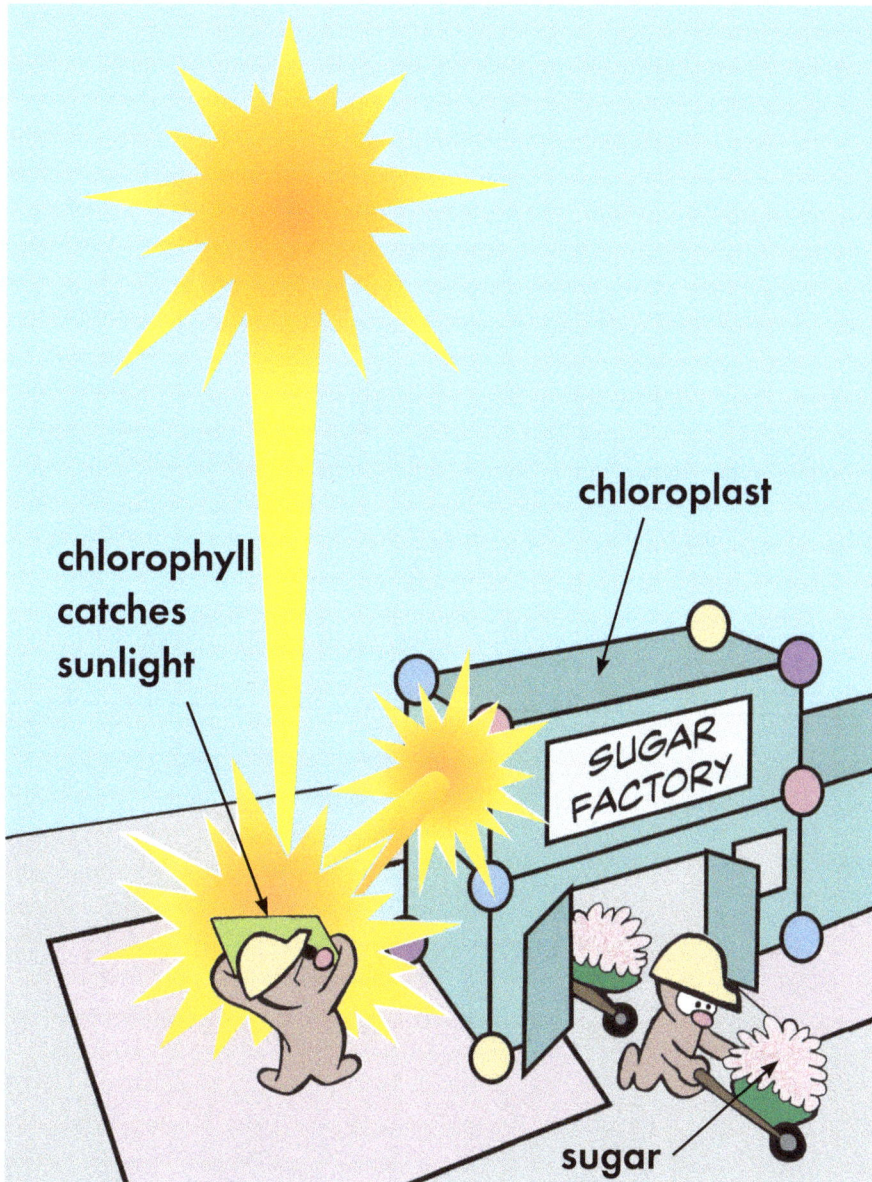

chlorophyll catches sunlight

chloroplast

SUGAR FACTORY

sugar

Chloroplasts are **organelles.**

Is cheese made in an organelle?

I don't think so.

Review: ORGANELLES

Organelles are small structures inside cells that do certain jobs.

Review: THE CELL

- All living things are made of **cells**.
- **CELLS** are made of **atoms** and **molecules.**
- Each **cell** has many parts that do different jobs.

Do plants use photosynthesis all year long?

Some plants, like evergreen trees, do use photosynthesis all year long.

But many plants stop using photosynthesis when the light is low in autumn.

Leaves change color in autumn when plants stop using photosynthesis.

In the spring, plants
grow new leaves, and
photosynthesis begins again.

How to say science words

atom   (AA-tum)

autumn   (AW-tum)

cell   (SEL)

chlorophyll   (KLAWR-uh-fil)

chloroplast   (KLAWR-uh-plast)

molecule   (MAH-lih-kyool)

organelle   (AWR-guh-nel)

photosynthesis   (foh-toh-SIN-thuh-suhs)

photo   (FOH-toh)

synthesis   (SIN-thuh-sis)

www.ingramcontent.com/pod-product-compliance
Lightning Source LLC
Chambersburg PA
CBHW040151200326
41520CB00028B/7566